YOUR THOUGHTS, YOUR REALITY

MYSTERIES BEHIND OUR THOUGHTS

Michael Lights

Unless otherwise indicated, all scripture quotations
are taken from the New International Version
Of the bible

Dedication

To all those who seek to fathom the depths of the mystery, our thoughts, despite the flaws in theories out there. Your relentless pursuit of understanding is an inspiration to us all.

Preface

In the profound journey of life, the mysteries of the human mind and its intricate connection to the spiritual realm have captivated seekers of truth for ages. "Your Thought, Your Reality" embarks on an illuminating exploration of this enigmatic relationship. As we delve into the pages of this book, we're invited to transcend the boundaries of conventional thinking and embrace a perspective that unites the tangible with the ethereal.

This preface serves as a prologue to the incredible voyage we're about to undertake. Together, we will unravel the profound implications of our thoughts, delve into the wisdom of ancient texts and spiritual teachings, and discover the transformative power of aligning our consciousness with the divine. Prepare to embark on a journey that transcends the boundaries of the ordinary, a journey where your thoughts hold the key to shaping not only your inner reality but the world around you.

Table of Content

Introduction

Welcome to "Your Thought, Your Reality," a captivating odyssey into the uncharted territories of the human mind and its intricate interplay with the spiritual dimensions. In this thought-provoking exploration, we embark on a quest to unlock the secrets of the mind's creative power and the profound influence it wields over our existence.

From the timeless wisdom of ancient scriptures to contemporary insights from neuroscience and metaphysics, we navigate a diverse landscape of knowledge. We'll traverse the realms of memory, consciousness, and perception, uncovering the profound truth that our thoughts are the architects of our reality.

Prepare to challenge your perceptions, expand your understanding, and embark on a transformative journey that holds the potential to reshape the world as you know it. Together, we'll embark on a path that bridges the gap between the physical and the spiritual, offering profound insights into the age-old adage: "As a man thinketh, so is he." This is a journey where

Chapter 1

Communication Beyond Language: Insights from the Spiritual Realm

In the spirit realm, communication transcends spoken languages and solely relies on thoughts and spiritual impressions. This is clearly emphasized in "Isaiah 65:24; Before they call I will answer, while they are still speaking I will hear". This is the reason why the devi, then Lucifer got expelled from heaven thus because of his thought. Note that as spiritual beings, our thoughts always precede our every move in life, and these thoughts are first sensed in the spirit realm before we even attempt acting physically upon them.

Spiritually sensitive individuals can effortlessly grasp the meaning and thoughts of one another, not through the knowledge of a specific language but through direct and telepathic understanding. It's important to note that God does not communicate in human languages, even though we often perceive God's messages in the languages familiar to us during our earthly life. No human language, no matter ho your thoughts become your reality, and your reality becomes your world.

Chapter 1

Communication Beyond Language: Insights from the Spiritual Realm

In the spirit realm, communication transcends spoken languages and solely relies on thoughts and spiritual impressions. This is clearly emphasized in "Isaiah 65:24; Before they call I will answer, while they are still speaking I will hear". This is the reason why the devi, then Lucifer got expelled from heaven thus because of his thought. Note that as spiritual beings, our thoughts always precede our every move in life, and these thoughts are first sensed in the spirit realm before we even attempt acting physically upon them.

Spiritually sensitive individuals can effortlessly grasp the meaning and thoughts of one another, not through the knowledge of a specific language but through direct and telepathic understanding. It's important to note that God does not communicate in human languages, even though we often perceive God's messages in the languages familiar to us during our earthly life. No human language, no matter how eloquent, can adequately express the Creator of all life. You can read more of this fact in my book titled "God's Ineffable Mystery".

When we hear God's message in a human language, it's because our spirits comprehend it in a manner that surpasses language and then convey it to our souls in a language we acquired during our earthly existence. Communication between spirits transcends language—it's akin to a profound connection (1 Corinthians 2:10, 11; Romans 8:26, 27). The reason why as humans we sometimes find it difficult retrieving certain information from our mind is because our mind is not fully tuned to the spirit realm but trust me when I tell you there is absolutely no way one can forget anything.

This form of communication is akin to an impartation of experience and knowledge or the transfer of data from one computer to another, although this analogy falls short in capturing the full depth of the experience and senses involved. One instantly "knows" (although this word is limiting because it involves experiencing the knowledge) what is being conveyed. As one progresses spiritually, the need for earthly language diminishes, gradually shedding the limitations of human language.

There exists a sense of knowing that extends beyond the physical world. In the spiritual realm, you pick up anything and desire to examine it, you seem to become one with that thing, gaining a comprehensive understanding because you've "experienced being a flower." Similarly, when you meet another being and wish to understand them, you seem to become that person and acquire comprehensive knowledge about them. However, this unique and supernatural form of "knowing" is only possible with the free-will consent of all parties involved. Remember this is mostly experienced in the prophetic ministry.

The Apostle Paul alludes to this realm when he speaks of knowing as we are known (1 Corinthians 13:12). This sense of being able to "become what you want to know" (without actually transforming into the object) represents an instantaneous infusion of knowledge encompassing not only objective facts but also subjective and perspective-based understanding. Consequently, everyone comprehends one another perfectly in the spirit realm. While spoken language is still used, it serves more for the beauty and diversity of communication and for specific purposes facilitated by the Holy Spirit's work. This is the reason why all of us are to make sure we leave a life of purity with the singular motive of pleasing and obey GOD the Lord.

No one can hide who they really are in the spirit realm. One is known as he truly is. There are no pretends, character disguise, identity theft, impersonation, no faking, you just name it. Everything one stands for are all boldly inscribed in one's spirit garment for all to see. There is nothing like privacy in the spirit realm. This is why the spirit realm beings always have an advantage over us and so they easily catch a glimpse of our thought informing them either to influence us positively or negatively depending on the quality of our thoughts, whether good or bad. Spirits in the spirit realm are very sensitive to the impulses of our thought, ever ready to influence and feed on our vulnerability to this realm. Always have a positive mindset.

The Path to Spiritual Consciousness: Connecting with the Spirit World

To truly comprehend and encounter the presence of God, a shift in our awareness is required. We need to transition from the domain of our soul and body to the realm of the spirit, as stated in "John 4:24". This transformation lies at the heart of meditation, a practice that allows us to establish a connection with the Spirit World.

Incorporating fundamental Christian rituals such as praying in the Spirit, worship, meditating on the Word, and prayer acts as pathways to attain this heightened state of spiritual awareness. However, it's not solely about achieving it; it's equally essential to sustain it. Maintaining continuous awareness of God's presence around the clock is entirely possible, as suggested in "1 Thessalonians 5:16-18, Galatians 6:18, 2 Corinthians 13:14, John 4:14, 14:16-18, 15:11, and 16:22".

As tripartite beings consisting of spirit, soul, and body, we possess the capacity to be conscious of three distinct realms, as expressed in 1 Thessalonians 5:23. The most elementary form of consciousness is body consciousness, characterized by thoughts consumed by the physical cravings and desires of the flesh. Those entrenched in this state are drawn away from the light and into the obscurity of spiritual dimensions.

Soul consciousness represents the second form of awareness, characterized by thoughts entwined with worldly matters. This realm breeds worry, anxiety, and contemplations centered on the transient aspects of life, all stemming from the soul. Jesus cautions against living in this manner, as noted in Matthew 6:25-34 and Philippians 4:6-7. Soul-conscious individuals often display selfish tendencies, prioritizing their own interests and revolving their world around their self-absorbed lives. When they depart from their physical bodies, they find themselves in the dimly lit spiritual realms. On occasion, individuals may exhibit a kind of "spiritual selfishness" aimed at their own lives or positions within ministry. This, however, is not genuine spirituality but rather a manifestation of soul consciousness parading as pseudo-spirituality. Those who fall into this category end up in the chilly, shadowed domains of the Spirit World.

The key to consistently savoring the presence of God lies in nurturing a state of mind/spirit consciousness. This form of consciousness is neither elusive nor restricted to ascetics and mystics; it is attainable by all. Anyone can achieve spiritual consciousness and tap into the flow of God's presence, which radiates from His throne and permeates all of His creation.

Occasionally, individuals may briefly enter this realm without conscious awareness. For instance, when you encounter a place of natural beauty that captivates you, or when you witness a breathtaking sunset or sunrise, or even when you take a leisurely stroll with your pet dog, shifting your thoughts momentarily away from worldly concerns to appreciate the companionship and scenery. During these moments, you unwittingly extend your perception beyond yourself and connect with the flow of God's presence and thoughts. For a brief span, you become cognizant of the melodious birds, the vivid hues of creation, the calming sounds of a flowing stream, and more.

Every aspect of creation is intricately designed to lead us to God, as highlighted in "Romans 1:20". This fleeting moment can evolve into your perpetual state of consciousness, even as you engage in your earthly responsibilities. This marks just the commencement of a journey filled with deeper levels of spiritual rapture, allowing you to walk in the complete embrace of God's boundless love.

TOP

Chapter 2

The Power of Divine Flow in Spiritual Consciousness

Sometimes, when you feel deep compassion and empathy for others, to the point of tears or overwhelming feelings of pure, unselfish love, you may be experiencing a moment of spirit consciousness. (The soul can also experience empathy, but it often carries a burden and selfishness, along with hidden emotions like anger, jealousy, indignation, and self-righteousness.) A moment of pure spirit-conscious empathy is marked by lightness and a sense of unity and peace infused with pure, holy love—the love that emanates from the God who is love (1 John 4:8).

During moments of human tragedy or need, those who step up to help can become engrossed in the desire and passion to assist others, sometimes achieving what might be considered "superhuman" feats. Unknowingly, they have tapped into the flow of spirit consciousness, and God's angels and ministering spirits assist them in performing extraordinary acts. Spirit consciousness can also be described as "love consciousness"—a love that is unselfish and sacrificial.

In the Spirit World, one's focus is not solely on oneself but is continually attuned to the love of God, the love present in all of His creation, and the love for others. There is an enduring sense of oneness with God and all of His creation while practicing the presence of God. As you learn, experience, and understand what spirit consciousness entails, you can integrate it into your daily life. If you encounter stress or strain during the process, pause, relax, and begin again. The subtleties of the soul may creep in. Remember that spirit consciousness involves letting go, yielding, resting, and uniting with God, while soul consciousness is marked by striving, stress, and tension. With time and patience, this state of consciousness becomes a part of your daily life, whether awake or asleep, as your spirit remains in communion with God.

Entering this rest signifies a release of your spirit from the soul and the body, even while you are physically alive, allowing it to become immersed in the divine flow of God's word, infused with life-giving energy (Hebrews 4:10, 12). By choosing spirit consciousness, all the intents and thoughts of your heart can align with the flow of life and thoughts from God.

Throughout our lives in the physical world, we are conditioned to be self-conscious (soul-conscious) and body-conscious. However, in the Spirit World, the opposite holds true: Stop thinking about yourself and start thinking about others. The presence of God isn't merely a self-centered, feel-good state. It is a state of being consumed by love for others, whether or not emotional feelings are present. Many people claim to have emotional feelings of love for God, but if this "love for God" does not translate into love for others, it often remains rooted in the realm of the soul (1 John 3:14; 4:7-8, 12, 16, 20-21).

While emotional states of compassion and empathy may be experienced as a result of loving God and others, these are side effects on the soul and should not be the main focus. Acknowledge the gift of emotions but understand that pure spirit and pure love transcend the realm of soul feeling; they exist in the realm of the spirit—a sense of unity with God. In the Spiritual World, we are taught to love others and be mindful of their needs. As we progress in loving others, we advance in the spiritual spheres.

Each day, immediately upon waking (and before sleeping as well—Psalm 4:4), take a moment to be still and listen to your heart. Engage in quiet meditation, glorifying God, and sincerely asking Him, with all your heart, to help you know Him better. Then, reflect God's love in your daily actions, thoughts, words, and prayers, showing kind love to make someone else's life happier. This daily exercise of love in your heart, mind, thoughts, and actions will gradually transform you from an inward-looking, selfish person into an outward-looking, loving individual who genuinely cares about helping those around you. This is the initial step toward opening your heart to the Spiritual World.

"Blessed are the pure in heart, for they shall see God" (Matthew 5:8). What is true purity of heart? It's when your heart aligns precisely with God's heart. Since God is love (1 John 4:8), you need to resonate at the same frequency of love to witness God in His manifestation. Your heart must beat in the same rhythm as God's, allowing you to see Him. The practice of spirit and love consciousness will enable God and His angels, who operate in the frequency of love, to make themselves known to you and guide you into deeper dimensions of the Spirit World.

Concluding this chapter, I would like to stress again that you put in every effort making sure all your desire and intentions towards life are pure, hence enhancing the pure nature of your thought in life which automatically will advance you massively in this life.

Now that I believe our minds are now prepared and set for the truth of the operations of the mind, with its consequences, whether good or good, let's now delve into the real deals of our subject matter.

TOP

Chapter 3

The Nature of Memories and Consciousness

You might have heard this many times and wondered why, as Christians, we are encouraged to change the way we think in order to understand God's perfect plan, which is the only path to repentance and renewal(Romans 12:2). Perhaps you've also come across the saying that a person is shaped by their thoughts (Proverbs 23:7). Here, I present to you the significance of strictly adhering this command and maintaining a continuous positive mindset, whether you're a believer or not. While I genuinely hope that everyone embraces the Lord Jesus Christ as their savior, I want to introduce you to Him if you haven't already, while also inviting you to explore this divine revelation from God's heart.

Being human, composed of Spirit, Body, and Soul, it's crucial to always bear in mind that each of these integral parts possesses a mind. Our Spirit has a mind, as does our Body, and naturally, our Soul also possesses a mind. As

Christians, we understand that, through the special Grace of the Lord Jesus Christ, one of these three components, the Spirit, has been redeemed. This Spirit stays connected to the spiritual realm and coexists with the Body and Soul. Of the two remaining components, the Soul has a dual role, responsible for both the Spirit and the material world simultaneously serving as "middle man or interpreter" between the our spirit and body and also between the spirit and physical world.

The source of all thoughts is in the realm of the spirit world, which our spirits are in contact with. Generally, memories are stored within the soul's realm of consciousness which is the mind of the soul, not solely within the physical brain of a human being. The brain is just one minute aspect of our mental experiences. When a human spirit departs its earthly body, it(spirit) retains all the memories from its earthly life. Just like having a dream. You realize that even though one may be asleep with all happening around his unconscious body resting on a bed, yet in the spirit realm life will still be on-ging in his dream But waking from sleep one is able to recall all the spiritual events because his physical brain start to receive these spiritual data from the mind of the soul which receive from the mind of the spirit. Even the cells within our physical bodies have their own centers of consciousness (Nerves). Our self-awareness originates from the consciousness of the soul.

When a person's spirit is in control among the three, their consciousness aligns with the spiritual realm. However, when there's discord between the spirit, soul, and body, conflicting states of consciousness arise within the soul, leading to inner conflicts, as described in "Romans 7:22-25".

We should prioritize allowing our spirit to guide our thoughts and consciousness over being solely influenced by the physical world, as emphasized in "Romans 8:5". It's important to distinguish between the thoughts and consciousness that originate from the spirit mind and those stemming from the soul's mind and body's mind, as mentioned in "Hebrews 4:12".

In the spiritual realm, memories, often referred to as thoughts, hold a deeper and more profound intensity compared to the physical world. Remembering an event in the spirit realm involves reliving the entire experience, not just a fleeting thought about it. Consequently, emotions like peace, love, and joy can be experienced on a vastly heightened level, but suffering can also be equally intensified.

In the spiritual realm, thoughts hold creative powers. As spiritual beings, every thought of ours emanates light, which serves as a revelation of our strengths, perfections, differentiations, and characteristics. Note that in the spiritual world there is only one source of light and that is what all spirit beings survive on and that is the light from the Son of God.

So Our Lord Jesus was not merely making a fanfare statement when he declared that He is the light of the world "John 8:12 and John 9:5". This spiritual light, flowing from God Himself through the Word (our Lord Jesus), is akin to the significance of blood in the physical world (Leviticus 17:11).

In the spiritual realm, life resides within this light(which is spiritual blood for better understanding) (John 1:4). Our thoughts on Earth emit either light or darkness based on our inner spiritual life of thought (Matthew 6:22, 23).

Our spoken words, which are just the vocal expressions of our thoughts, align with the light of our thoughts hence "out of the abundance of the heart(mind) the mouth speaks (Luke 6:45). Just as physical blood can be analyzed to diagnose the state of the physical body, these lights emanating from our spiritual life through our thought reveals the strengths or weaknesses of our spirit.

Thoughts originating in the realm of spiritual light can sometimes be sensed by those attuned to them on Earth. This is how prophets and other men of God are able to perceive and tell the thoughts of people. They radiate and inspire earthly thoughts to higher levels through the Holy Spirit. Similarly, the thoughts of malevolent spirits residing in darkness can influence those who are drawn toward darker inclinations. The extent of this influence depends on the inner strength and character of each individual's heart.

Like attracts like—light attracts light, and darkness attracts darkness. It is the free choice of every individual to which realm of thought influence they yield. The impact of thoughts from the spiritual realm on the physical Earth is more significant than many realize. When combined with the actions of angels and saints, as opposed to the work of evil spirits, these thoughts shape the activities on Earth.

The invisible world continues to exert an astonishingly great influence on the physical world. The good and holy thoughts of Earth's inhabitants are visible in the spiritual realm as radiant light in various colors, offering insight into their inner natures. This luminance, a reflection of our thought character, can unveil our strengths and weaknesses. Through it, God and angelic spirits can discern areas where we need assistance and areas that require transformation.

When we depart from our physical bodies and dwell in the spirit world, our entire spiritual being will emit light, exposing our strengths and weaknesses. In the spirit world, no one can conceal their true nature, as it is plainly visible to all through the light they radiate. The dominant aspects of our character are mirrored in our spiritual attire, classifying us within specific spheres.

TOP

Chapter 4

Our Thoughts Free Will Choices and Consequences

Almost all of the things that happen to a human life are through their own free will no matter how many external causes or currents are taking place in the environment into which they are born. Not just in the sense of a fleeting thought which impinges upon a mind but rather a thought that is accepted and becomes part of the soul life and thought pattern of the individual life.

Free Will

All actions have a cause and effect – a sowing and reaping for its doer (besides affecting the lives of those around them). All thoughts that are accepted into one's thought pattern and character attract forces that control the circumstances around the individual life. Dark thoughts attract dark forces and bright thoughts attract good forces. Thus the real battle in life is to choose both the right actions and the right thoughts (2 Corinthians 10:3-6). Both actions and thoughts have a cause and effect in this life and in the next life.

Those who do not yield to the forces of good and love thoughts become destroyed by forces of darkness that are attracted to the darkness they allow in their hearts. All humans are like magnets that constantly (by their free choice of thought) attract to themselves forces of good or bad which work for them or against them. No matter how weak they think they are, by choosing the right thought in their heart of hearts, they release forces of good that would gradually overcome all the power of evil assailing against them. One small loving thought or action can fell a tsunami of evil forces. Evil spirits are powerless against the love of God which is thought of good and love. Perfect love casts out fear (1 John 4:18).

When a person is influenced to do evil or wrong things, it is because they attract evil spirits to themselves by their own evil thoughts. To be tempted is not wrong but to entertain thoughts of temptation which are evils and allow it to produce an evil desire in our lives is wrong. This evil desire attracts evil spirits who then entice us to do evil. The thought of good or bad is our choice, the enticing is the encouragement received from evil spirits to do evil. Each one is tempted when he is drawn away by his own desires and enticed (James 1:14)

Wrong emotions and thoughts of hate, lust and anger stir up and attract evil spirits like filth attracting flies (Ephesians 4:26, 27; Matthew 5:22). Passionate agape love is like a bright light in the spirit man that lights up the darkness and draws ministering spirits together to the service of the individual while dissipating evil spirits. The angel of the Lord encamps all around those who fear the Lord (Psalm 34:7). One should be careful of allowing negative emotional thoughts to dominate one's life. Only the fruit of love, joy, peace, longsuffering, gentleness, goodness, faithfulness, meekness and temperance should be allowed to remain in our character (Galatians 5:22, 23). It is clearly visible in the spiritual world whether one is cold, or lukewarm or hot and has the first love for the Lord Jesus (Revelation 2:4; 3:16)

We should cultivate positive emotions and thoughts in line with the fruit of the Spirit by being passionate for the things of the Holy Spirit. Many people think that yielding to the Spirit of God is just a passive matter of letting go and doing nothing. The Holy Spirit will not work without some contribution of free will from us. We need to desire the things of the Spirit. When our desire for the things of the Spirit and of God becomes overwhelmingly high above all other desires, when our passion and love is undivided for God and God alone, there is a spiritual union that takes place in the bond of the Spirit with our spirits such that the energy of the Holy Spirit, and all the might of the work of God's angels guides our lives and destinies.

Thus it is important never to allow wrong or evil thought to form in our heart – they attract evil spirits to our lives who will cause us to sin. The free choice is still ours. We must stop sin at its temptation stage and not allow it to grow into the desired stage. When our heart is pure and no
wrong thoughts exist, not a single evil spirit can stand in our presence. Angels and all the heavenly hosts are attracted to our lives and encamped around us when we have a pure

heart. A pure heart of love is the greatest asset in the spiritual world. Blessed are the pure in heart for they shall see God (Matthew 5:8).

TOP

Conclusion

The most important part of your life is the ability to use your mind the right way. Maybe you didn't know that but now you do. Your thoughts define your life. They define your value. They define your personality. You can never be better than your thoughts. You are no greater than your thought because how you think defines you. The character of your thoughts is the character of your personality.

Watch your thoughts. No man ever got better than his thoughts. Your very life is the expression and manifestation of your thoughts. It doesn't take long for people to know who you really are. All you need to do to expose your thoughts is speak out for a few moments, and they know how you think and them knowing how you think enables them to know who you are.

Think good, straight and pure about people, nature and even yourself. Never think failure, defeat, weakness, alone, rejected, unlucky. Always be positive about all things. Use your creative power well.

TOP

www.ingramcontent.com/pod-product-compliance
Lightning Source LLC
Chambersburg PA
CBHW072227290526
45794CB00007B/2919